201 HANDY HINTS FOR HORSE PERSONS

201 HANDY HINTS FOR HORSE PERSONS

KAREN BUSH

Illustrated by Clare Colvin

KENILWORTH PRESS

First published in Great Britain by
The Kenilworth Press Ltd
Addington
Buckingham MK18 2JR

Reprinted 1991, 1992, 1993, 1994
Hardback edition 1997

British Library Cataloguing in Publication Data
A catalogue record for this book is available
from the British Library

ISBN 1-872119-00-X hardback
(ISBN 0-901366-65-X paperback)

Printed in Great Britain by Hillman Printers (Frome) Ltd

Introduction

More and more people are keeping horses and ponies of their own these days, and though you do not need to be tremendously wealthy in order to do so, it can still prove to be an expensive and time-consuming hobby. Obviously you should never consider attempting to economise on things which are essential to your horse's wellbeing, such as regular feeding, worming, shoeing and veterinary attention, but there are still a number of ways in which it is possible to save both time and money.

If you are new to horse or pony owning, you probably have not yet found out about many of these legitimate 'short cuts'. Even if you are an experienced owner, it does not mean to say that you have stumbled across all of them. You will find that you keep discovering new ideas all the time – for as long as you continue to keep a horse in fact. This book tries to give you at least a head start if you are a novice, or to nudge your memory if you are more knowledgeable, by setting down some practical hints and tips which will make life a little easier and your stable management more successful.

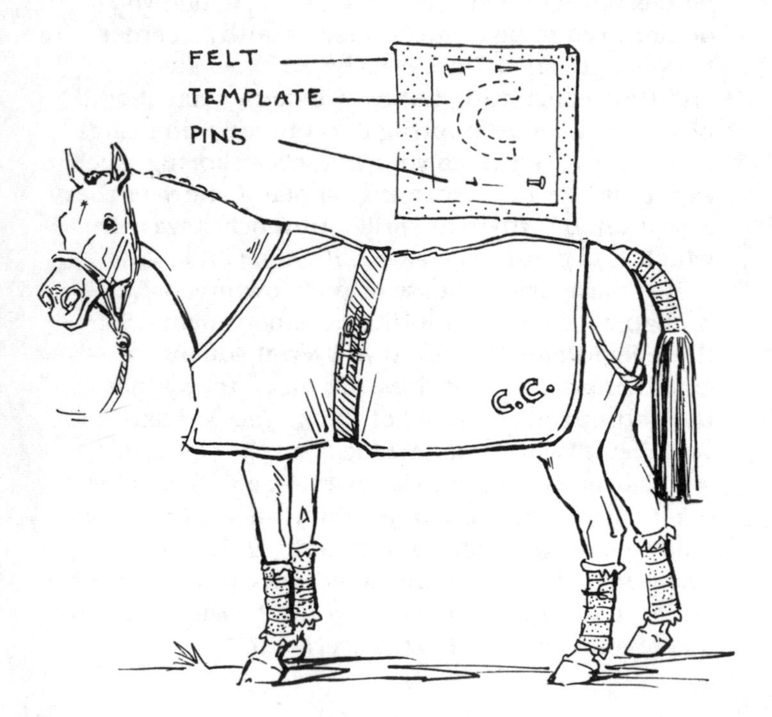

Personalise your own rugs for shows and smart occasions.

1

When trimming your horse's whiskers, save a little time and produce a neater result by using a plastic safety razor instead of scissors. It is also safer if he tends to fidget.

2

The seams of New Zealand rugs often leak. Waterproof them effectively and cheaply by rubbing a piece of candlewax along the stitching.

3

To produce a good shine on rubber boots, use a little furniture polish from an aerosol.

4

To give a more professional touch to your rugs, add your initials to the nearside corner. Either buy the letters individually (from saddlers, needlework or craft shops) or make your own by drawing a paper template, pinning it on to a square of felt, and cutting around it. You can then stitch it into position.

5

If you keep losing your hoofpick in your horse's bedding, plait up a piece of brightly coloured baler-twine and tie it to the handle.

6

To make a white tail look really white again, wash it using a *little* biological soap powder. **Warning:** Be careful not to get it on to the skin of the dock, as it may irritate and cause rubbing.

7

When your velveteen-covered riding hat becomes faded and scruffy looking, smarten it up by using an appropriately coloured suede shoe spray.

8

To make a cheap feed scoop cut diagonally across an empty plastic squash bottle.

Do-it-yourself feed scoops.

9

Use a piece of baler twine to hang up a penknife or a pair of scissors near to your supply of hay or straw, ready to cut open the bales.

10

Really determined 'jockeys' (blobs) of grease on tack can be removed either with a fingernail or a bunch of horsehair.

11

A plastic washing-up bowl makes a cheap and useful feed bowl which is also easy to clean.

12

Trimming off the corners of your number at shows, so that they are slightly rounded, will make the number look smaller and neater, and will also prevent it from curling up.

13

Tie a knot in the end of your lead rope when leading your horse, so that it does not slip through your hands. Wear gloves, too, so that if he tries to pull away you have a better grip and thus avoid burning the palms of your hands.

14

Remove bot fly eggs with a disposable plastic safety razor. This is far cheaper than buying a bot knife.

15

If your horse has a very greasy coat which spoils his appearance or makes him difficult to clip, tackle the worst areas by adding a little Dettol to a bucketful of warm water. Wring a sponge out in it, and then rub it vigorously against the lie of the hair. Rinse it out frequently. You will see the grease beginning to form a scum on the top of the water as it comes out of the coat.

16

Use old saddle soap tins as pegs for bridles. Just nail them to the tack room wall. A rounded peg such as this will help to keep the headpiece in shape as well as keeping the bridle tidy.

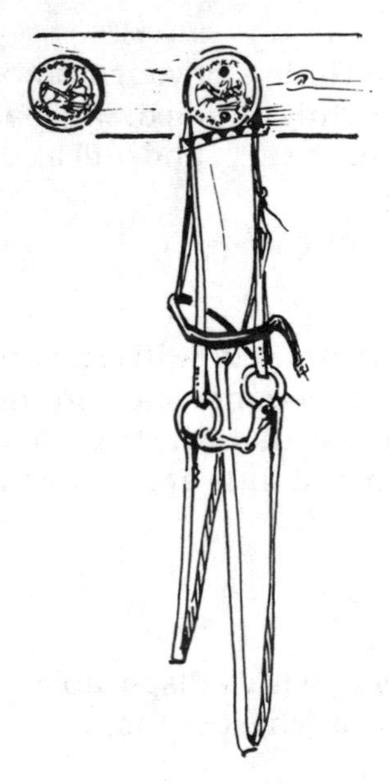

Old saddle soap tins make handy bridle pegs.

17

If your horse scrapes the ground in front of his stable door with a forefoot, place a rubber mat there. It will prevent the toes of his shoes from being worn out quickly and if he cannot hear the noise, he may stop. Try to turn him out as often as possible, too, in case the habit is caused by boredom.

18

Make your own hoof oil by mixing a little Stockholm tar with cheap vegetable cooking oil.

19

Make some extra money towards paying for field rent or feed by bagging up and selling muck from your stable; this will also keep your muckheap to a reasonable size.

20

Save old pieces of saddle soap which are too small to use and, when you have enough, heat them gently in an old saucepan until they melt. Pour the liquid into an empty tin or margarine container and leave it in the fridge until it sets.

21

Put worm powders into the refrigerator a day or so before you want to give them to your horse. It reduces the smell so that they are not as off-putting.

22

Blanket-stitch around the edges of pieces of Gamgee which are used beneath bandages. It will stop them from fraying and they will last longer.

23

It is cheaper to make your own haynets out of baler-twine than to buy them. About 24 strings will make an average-sized one. Tie them all into a large knot, which can be hung from a nail. Then knot alternate pairs of strings together to complete one row. Knot pairs of strings together on the next row in a similar way, but using one each of the strings from the pairs knotted in the first row, so that large holes are formed. You can make these holes bigger or smaller, depending upon the distance apart that you make the knots. Continue in this way until the net is complete. When you have finished, heat the knot to seal the strings together and plait up some baler-twine to make a drawstring.

How to make a baler-twine haynet.

First row (upside down)

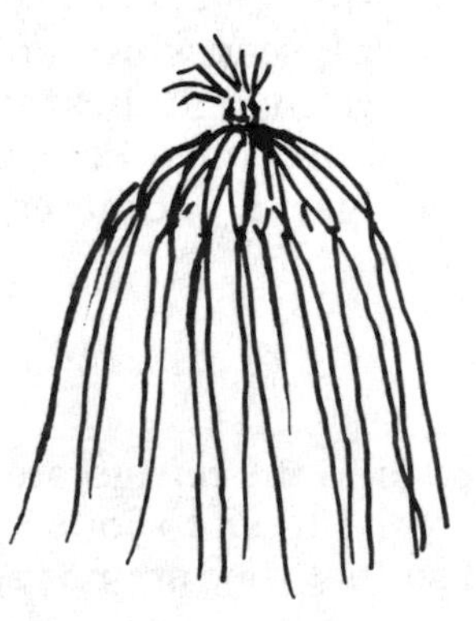

Second/third rows

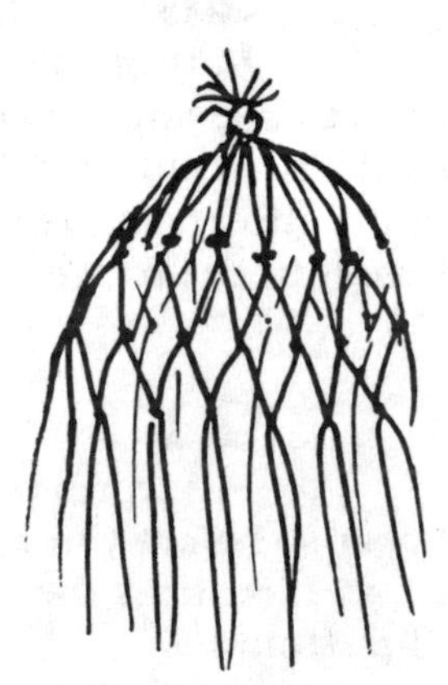

Thread the drawstring *Filled haynet (right way up)*

24

Make your own dressage markers by collecting plastic cones or old oil drums and tins, and then painting the letters on them and using them to mark out a 20m x 40m working area in your field.

25

Before going into the show ring dab a little baby oil around the eyes, muzzle, on the chestnuts and beneath the tail to enhance your horse's appearance. Leave this until the last moment, so that hairs and dust do not stick to the oil.

26

If New Zealand rug leg straps tend to chafe, and if you have ensured that they are kept clean and supple and are not too tight, try slipping a piece of bicycle inner tubing over them.

27

The cheapest antiseptic wash for cuts is salty water. Never use Dettol or disinfectants for cleaning wounds, as they will impede the healing process.

28

As a cheap fly repellent use vinegar.

Vinegar is a cheap and effective fly repellent.

29

Never tie your horse directly to a fixed object. Always tie his lead rope to a breakable piece

of string, which will snap if he panics and pulls back, thus saving his headcollar and neck from damage.

30

When washing girths in the washing machine, place them all inside an old pillowcase to stop the buckles from doing any damage.

31

Put a squeezy rubber ball in the field water-trough in winter; it will help to prevent it freezing over completely, and will leave a drinking hole in it until you arrive to break the rest of the ice.

32

When your old lead rope wears out, make a new one by plaiting some baler-twine and attaching it to the old clip.

33

Put a tail bandage on your horse's tail while he is being clipped, to keep the hairs safely out of the way of the clippers.

34

Put a few drops of baby oil on a soft brush, and carefully brush through the tail while it is clean. As well as helping to stop tangles, the oil will also make the hairs less brittle and will help to prevent white hairs becoming discoloured.

35

If scissors or a knife are not to hand, you can easily cut through the twine on a bale of hay or straw by using another piece of twine. Slip it beneath the strings and, holding an end in each hand, use a sawing motion to cut them through.

Use a piece of twine to cut through another.

36

An old sleeping bag can easily be converted into a warm quilted stable rug for the winter. Remove the zip and cut out a semi-circular section at one end for the neck and shoulders. Stitch around the edges, adding a coloured binding if you wish. Use strips of broad Velcro for breast straps.

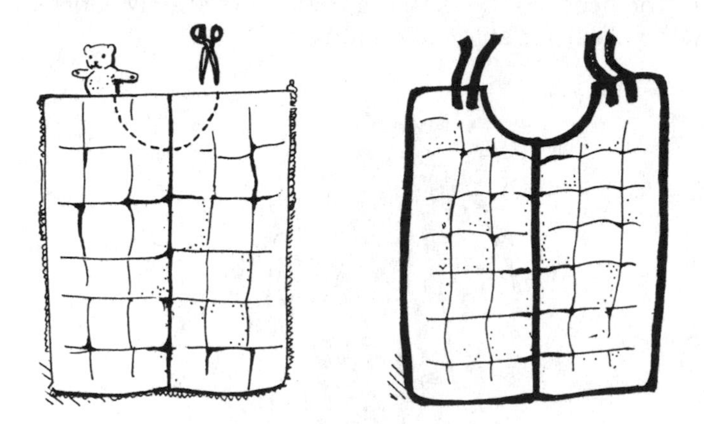

An old sleeping bag can be converted into a smart quilted rug.

37

Flies are attracted more to sweating and dirty horses than to clean ones, so bathe your horse regularly during the summer months, and if the weather is warm give him a quick sponge down after exercise. Shampoos with a mild antiseptic added also act as an additional protection against flies.

38

After an injury has healed up, the hair may not grow back in the affected area for some time.

Stimulate regrowth by rubbing either a little Vaseline or Cornucrescine into the skin.

39

If your horse's forelock is a bit sparse, make him a fly fringe to keep flies away from his eyes. Loop pieces of string around an old browband and attach it to the headcollar, ensuring that it fits snugly. Check it regularly for signs of chafing.

A home-made fly fringe.

40

Crib-biters can be deterred by mixing up some mustard and smearing it on projecting surfaces.

41

Use an old tea towel as a stable rubber. It is cheaper than buying the real thing from a tack shop.

42

Dark-coloured jodhpurs for everyday wear show the dirt less and will not need washing as often as light ones.

43

In showing classes, use cotton thread of the same colour as your jacket to attach your number: it looks neater.

44

Cheaper than coat gloss for a gleaming finish on your horse's coat is a little aerosol furniture polish sprayed on to a stable rubber and then wiped on. **Warning:** try a small patch first just to check that there are no allergic reactions.

45

Remove all the grease and dirt from your tack by using hot water with a little washing-up liquid added to it. Soap well afterwards, and oil occasionally so that it remains supple.

46

Surcingles are cheaper to buy than rollers. To prevent a surcingle from rubbing or pressing on the back, place a thick piece of sponge beneath it at the point where it crosses the spine.

A piece of thick foam rubber placed beneath a surcingle will help to prevent direct pressure on the spine.

47

When applying louse powder, work it all into the roots, starting at the forelock, going down the crest, along the spine and into the roots of the tail. Sprinkle it on both sides of the mane, not just along the top, so that you do not miss any.

48

When washing your horse it is cheaper to use a family-size brand of shampoo from a supermarket than a special equine preparation – but try a small test area first to make sure that there is no allergic reaction.

49

If your horse keeps kicking his water bucket over in his stable, put it in the centre of an old car tyre.

An old car tyre makes an ideal bucket holder.

50

Plait up some baler-twine and hang it across your tack room or feed shed, so that you can put wet rugs on it to dry out.

51

An old plastic laundry basket makes a good skep.

52

If your horse is difficult to catch, try keeping a piece of rustly paper in your pocket to arouse his interest and to lure him close enough to be tempted by the carrot or pony nuts you have brought for him.

53

If you are going to a show, shampoo your horse's coat a day or so beforehand so that it has time to settle and regain its natural sheen for the big day. A mane which has been recently shampooed is difficult to plait up.

Shampoo your horse a day or so before a show to give his coat a chance to settle down.

54

Brittle feet can be improved by adding a sachet of gelatine dissolved in a mugful of hot water to one of the feeds each day.

55

If you buy secondhand saddlery or horse clothing, disinfect it thoroughly before use, using a strong solution of Milton Sterilising Fluid (available from chemists). If your tack should ever become mouldy, disinfect it before using it again, since the fungal spores will not be destroyed by normal cleaning, and can set up a skin infection.

56

If you have to get up very early to go to a show, if may be more convenient to plait your horse's mane the night before. To keep the finished plaits neat and clean, put a tissue over each and secure it with an elastic band. Alternatively, use one leg from a pair of tights, running it along the length of the neck, and securing it over each plait with an elastic band. Always leave the forelock unplaited until the last moment – the horse might rub it during the night.

57

If your horse tends to finish off his hay very quickly, buy or make him a haynet with smaller holes.

58

Provide some amusement for your horse when he is stabled by boring a hole through a large turnip. Thread a piece of rope or plaited baler-twine through this and hang it up in the stable for him to nibble at.

A turnip hung up in the stable can help to prevent boredom.

59

Shoe whitener is effective on white socks when you are entering a showing class. Alternatively, use a block of white dog chalk, or even talcum powder, to whiten them.

60

If you find that your whip tends to slide through your hand because there is no knob on the handle, fix a rubber martingale stop to it.

61

Sheepskin numnahs (natural, not man-made) can be kept reasonably clean if you sprinkle them with a little talcum powder after use, and then gently brush it out. It will absorb some of the sweat and dirt.

62

Dark coloured gloves are preferable to light or white ones when competing in dressage or showing classes: they draw less attention to the hands.

63

Linseed helps to promote a healthy bloom on the coat, since it is rich in fats. It is easily digested and will help to put on weight. To prepare a jelly, use 20 parts of water to one of linseed. Soak for six hours, strain off the water and replace it with fresh water. Using a heavy saucepan, bring it to the boil and simmer uncovered for four hours, stirring occasionally. This will form a thick, jelly-like substance, a teacupful of which may be fed two or three times a week. Before feeding, check that the seeds have broken. **Never** feed uncooked linseed: in their raw state the seeds are poisonous.

64

Clean and keep old rugs instead of throwing them out. They can be cut up and used to repair tears in newer rugs.

65

The way in which you plait a mane can do a lot to improve the appearance of your horse. Lots of

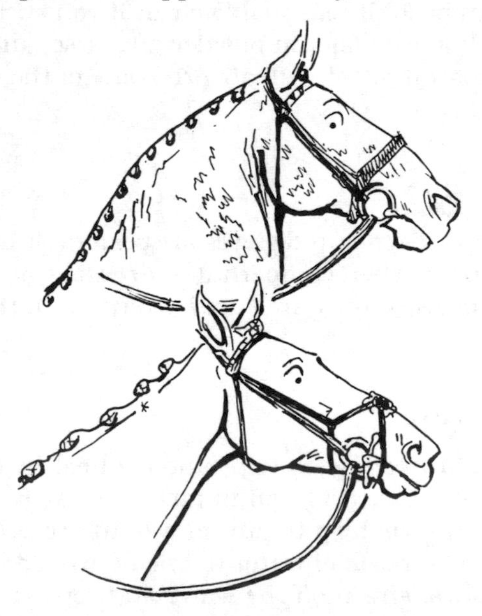

The way in which you plait a mane can deceive the eye and enhance your horse's appearance.

small, dainty plaits will make a short, thick neck look better proportioned, whilst fewer, larger plaits built up along the crest will make a long, thin neck appear to carry more muscle.

66

When turning a horse in hand, always turn him away from you so that he does not tread on your toes. If you are showing him off for a vet or in a showing class, it also means that you will not block the view of whoever is watching.

67

Trimming away a small section of mane just behind the ear makes the bridle headpiece sit better.

Trim away a little of the mane behind the ears so that the bridle sits comfortably and to make it easier to put in the first plait.

68

Codliver oil is cheaper to buy in a large gallon tin than a small bottle. If you want to use it as a feed additive to improve the quality of the coat, split the cost, and the oil, with a friend.

69

Leave feathers untrimmed during the wet winter months, as they will act as natural drainpipes, leading water away from the heel region and helping to prevent cracked heels. The heels can be further waterproofed by smearing a little Vaseline in them.

70

Billets are more easily undone by pushing the leather up into a loop so that it slides off the hook stud. It also prevents the leather in the area becoming damaged and weakened – and rules out the likelihood of breaking your nails.

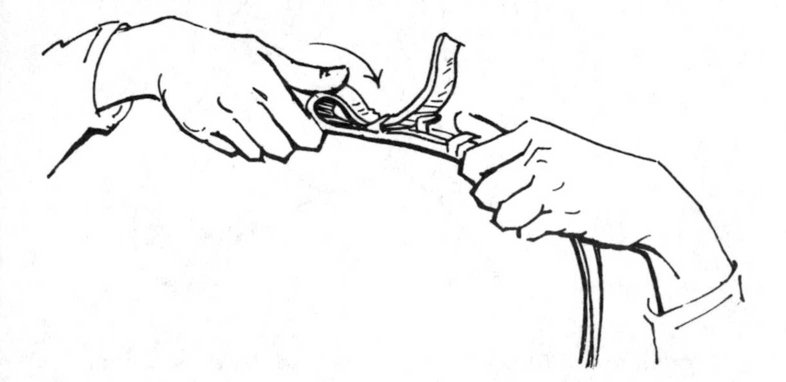

Undo billets correctly to avoid damaging the leather.

71

Clean your grooming kit at the same time as you shampoo your horse, so that you do not brush dirt straight back into his coat.

72

Dusty cobwebs will make a box dull, may fuse light bulbs and can aggravate stable coughs. They should be cleared away regularly.

73

Overreach boots can sometimes cause chafing if the horse is unaccustomed to wearing them. Smear a little Vaseline around the tops so that they slide freely around the pastern.

74

Change the type of wormer you use once a year – worms can build up a degree of immunity and a wormer may become less effective after prolonged use. Do not simply change the brand name, but check the ingredients on the label, to make sure that the wormer contains a different anthelmintic.

75

Tie a hoofpick to an old bucket by using a long piece of string. This will stop you from losing it, and will provide you with a container into which you can pick out the feet so that you do not mess up the bed or stable yard.

76

When trimming the end of a tail, ask someone to place an arm beneath it, so that it is in the position in which the horse carries it when moving. Otherwise you may end up cutting it too short.

To avoid cutting your horse's tail too short, ask a friend to hold it at the height the horse carries it when on the move.

77

If you do not have, or cannot afford, a shovel, scoop up piles of droppings between two pieces of board measuring about 9 in by 12 in. Piles of straw or hay can also be picked up in this way after they have been swept up.

78

To make an interesting and solid looking jump drape a piece of old carpet or fake 'grass' begged from the greengrocers over straw bales or a pole.

79

Oil the leather on lampwick girths and on rugs before washing them, or apply a little Vaseline, so that the water and detergent do not do any damage. Afterwards, wash and soap the leather thoroughly.

80

Change your stirrup leathers over from one side of the saddle to the other once a week. This will prevent the nearside one from becoming stretched by being used for mounting all the time.

81

If your horse is a fussy feeder, medicines can easily be given by placing them inside a hollowed-out apple or carrot. Alternatively, mix powders (tablets and pellets can be crushed) with black treacle and smear it on the tongue with a wooden spoon, so that it cannot be spat out again.

82

Make a check list of everything you will need to take with you to a show and tick the items off as they are loaded into the car or trailer so that you do not forget anything.

83

Baler twine can be used as a sweat scraper by taking a doubled-up length and drawing it down across a wet coat.

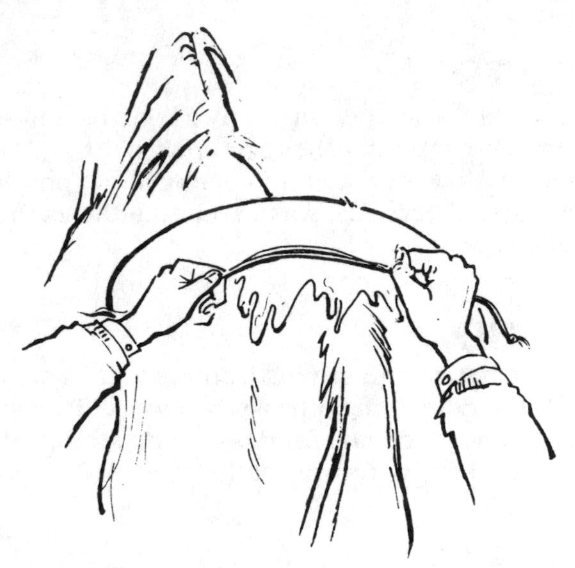

A doubled length of baler-twine makes a good sweat scraper.

84

Water in the stable should be changed regularly. It absorbs ammonia from the stable atmosphere, which gives it an unpleasant brackish taste that will put your horse off drinking.

85

Three or four times a year check that your saddle fits correctly. Remember that your horse's shape will change according to the quality of grazing and the amount of exercise he is getting.

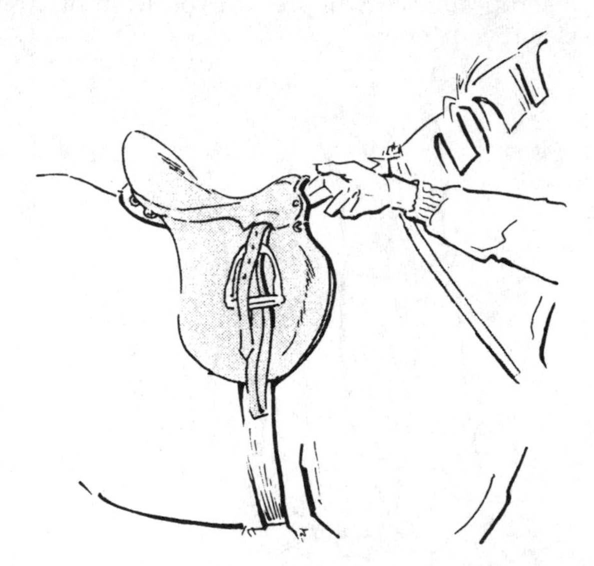

Horses may change shape during the year: check regularly that your tack fits correctly.

86

Cubed and coarse feeds have added supplements. Giving an additional one can imbalance the ration and is a waste of money.

87

If your horse grazes out with others, arrange with the owners to worm them all at the same time. If you worm yours separately it will not be very effective, as he will simply ingest eggs and larvae expelled by the other horses.

88

A boot jack is very easy to make out of a piece of scrap wood. It will prevent you from pulling the heels off your boots.

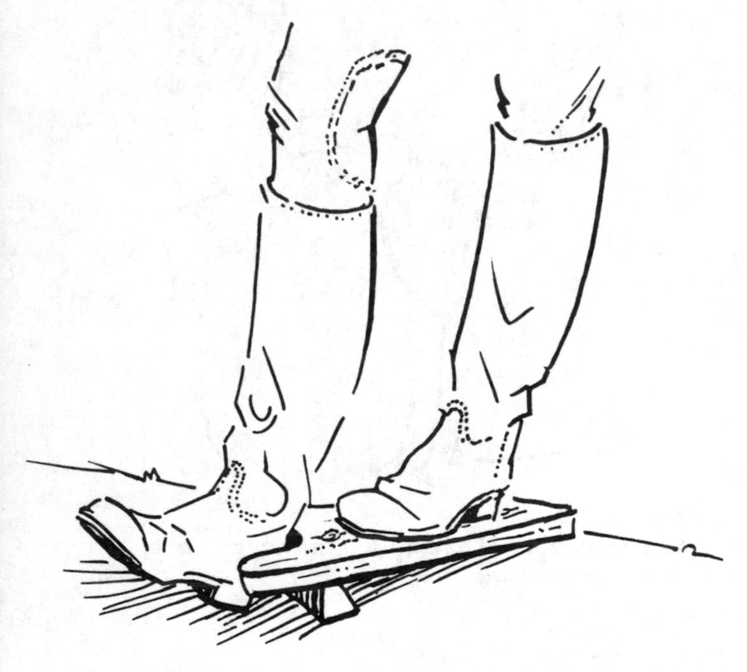

A boot jack is easy to make and saves the heels of your boots.

89

It is a false economy to feed cheap, inferior feedstuffs. Because their nutritional value is likely to be low, large quantities have to be given, so you are not really saving money. If they are dusty or contain fungal spores they may also be harmful to your horse's health.

90

Avoid making saddle soap too wet by spitting on it. If it becomes too lathery it does not help the leather.

91

Never leave your horse standing in the stable without any bedding. If he is likely to get his legs splashed, he may refuse to stale and that can lead to kidney troubles.

92

Keep lead ropes tidy when they are not in use by making a loop just beneath the clip and winding the remainder of the length of rope around it. The end can then be tucked through the bottom of the loop to stop it from coming undone.

Keep lead ropes tidy.

93

Shavings can be an expensive form of bedding, so keep wastage to a minimum by placing soiled bedding in a wire shopping basket. The clean shavings can be shaken back on to the bed, leaving the droppings in the basket.

94

Buy a pair of rubber gloves for skepping out droppings. Hands are quicker and more economical than shovels.

95

Wrap chicken wire around trees in fields to stop horses from chewing the bark – but make sure that it is properly secured to prevent injuries.

96

If hay needs to be fed wet, the best way of soaking it is to fill a hay net and place it in a trough or dustbin full of water overnight.

97

Ties to go with show outfits can be bought cheaply in jumble sales and secondhand shops.

98

An easy way of keeping feathers trimmed and neat is to use a trimming comb of the type used for dog grooming, which has a replaceable razor blade screwed in over the teeth.

99

Horses who try to barge out of their stables can be foiled by fixing a breast bar across the doorway on the inside. It also allows you to duck under and go in and out of the stable while you are grooming or mucking out, without having to open and close the door all the time. During the summer it will also keep the stable cooler.

Breast bars keep stables cool in summer and stop ill-mannered horses from barging out when you open the door.

100

Hang lampwick girths up to dry so that the leather parts are at the top and water does not drip on to them.

101

A wire-bristled dog grooming comb is very effective for removing dried mud and for getting the loose hairs out of winter coats. Do not buy one which is too scratchy, and do not use it on delicate areas where the skin is sensitive.

102

Pick the feet out before taking your horse out of the stable – it saves having to sweep the yard again afterwards.

103

Put the saddle on from the offside to save having to walk around your horse in order to let the girth down and to check that it is not twisted.

104

If your horse returns wet from a ride during the winter, put him in his stable with plenty of dry clean straw beneath his rug, which should be turned upside down to stop the lining from becoming saturated. Secure everything with a roller or surcingle. The straw will allow air to circulate, thus hastening the drying off process, and will keep him warm so that he does not become chilled.

105

When you are plaiting up before a show, save time by threading lots of needles beforehand. Stick them into the front of your sweater so that you do not lose them and they are handy to reach.

106

Always keep a list of useful numbers by your telephone, including those of the vet, farrier, saddler and feed merchant. It is quicker in an emergency than going through the telephone directory.

Keep a list of useful numbers close by the phone.

107

Bicycle puncture repair kits can be used for repairing rubber riding boots if you accidentally push a pitchfork through the toe while mucking out. The boots will not be smart enough for shows, but the patches will at least stop leaks and will keep your feet dry around the yard.

108

If you wear long rubber boots and find that they tend to slip downwards (especially when riding without stirrups) try wearing an extra pair of thick socks. If there is not enough room in the foot of

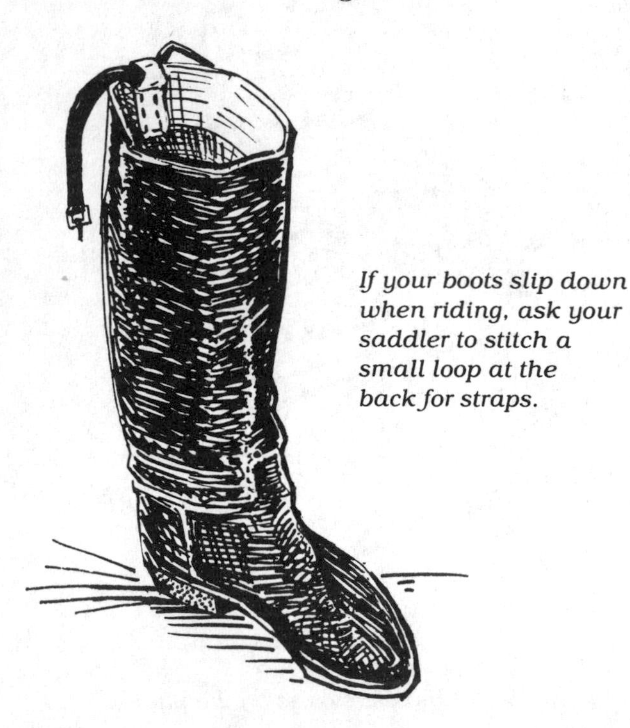

If your boots slip down when riding, ask your saddler to stitch a small loop at the back for straps.

the boot to allow this, leg warmers will do instead. You could also stitch a small leather loop on to the insides, at the back seam. Slip an old spur strap, or real boot straps bought from the saddlers, through each one; they will not only help to prevent the boots slipping down, but will also make them look more like leather ones for showing.

109

Wrap some Sellotape, sticky side outwards, around your hand and use it to remove hairs from your show clothes and numnahs.

110

A seam unpicker from a needlework shop is useful for quickly undoing plaits without accidentally cutting off pieces of the mane.

111

If your horse's surcingle or roller tends to slide backwards, either attach a breast girth to keep it in place, or ask your saddler to make some reinforced slots on either side of the rug, through which it can be slipped.

112

To prevent a horse eating his feed too quickly put a large lump of rock salt in his feed bowl. This will also ensure that he receives as much or as little salt in his diet as he wants.

113

After tightening your girth, and before mounting, always pull each foreleg forwards to ensure that the skin behind the elbow does not become pinched.

Before mounting, stretch each foreleg forward to prevent the girth from pinching.

114

Horses with scurfy coats may be treated with an infusion of rosemary, which should be made up as follows: add approximately $1/2$ pint of water to 1 tablespoonful of dried rosemary. Bring it to boiling point, cover, and simmer for 2 minutes. Remove from the heat and, keeping covered, leave for 5 to 6 hours or overnight. Strain off the rosemary, and use the liquid as a rinse after shampooing. Check that any scurfiness is not due in the first place to not having thoroughly rinsed soap out of the coat after bathing.

115

Use boot polish instead of hoof oil if you are showing your horse in an indoor arena. The shavings will not stick to it.

116

New Zealand rugs often have surcingles attached to keep the rug in place. Prevent them from rubbing by stitching the surcingle to the rug

Stitch surcingles on to New Zealand rugs, leaving a loop where they pass over the spine.

on either side of the spine, leaving a loop so that pressure is not placed on the backbone.

117

Always remove your gloves, however cold the weather, when picking out feet. It is a good opportunity to detect any signs of heat or swelling in the legs.

118

A few stalks of elder stuck through the browband will help to keep flies at bay while you are riding. The smell of the crushed stalks repels insects.

119

For a horse who is not very fit and is being brought back into work after a rest, a sheepskin sleeve fitted round the girth will prevent chafing. It can easily be made from a piece of real or synthetic sheepskin. Cut out an oblong, stitch it into a tube shape and slip it over the girth. To reduce the danger of sores or galls always keep tack well cleaned and supple, numnahs and girth sleeves regularly washed, and the horse well groomed.

120

Plait the end of a wispy tail while it is still wet after washing, and secure the end with an elastic band. When it is dry, unplait it and comb it through with your fingers. This will give it a slightly wavy and thicker appearance.

121

To keep the tail out of the way whilst you are trimming up or bandaging hind legs, tie a knot in the end of it.

Tie a knot in your horse's tail to prevent it getting in your way when you are working near his hind legs.

122

A chequer-board pattern can be made on the quarters for shows if you have a good eye and a piece of broken comb. Wipe a dampened cloth over the quarters in the same direction as the hair, and then use a piece of fine-toothed comb in a downwards movement across the lie of the hair to make each square.

123

Use a clothes peg or bulldog clip to peg the rest of the mane out of the way whilst you are plaiting.

124

To prevent breaking the hairs off short, never brush out a wispy tail except when it is clean (i.e. just after washing). The rest of the time just tease tangles out with your fingers.

125

Wear an old shirt or nylon overall over your jumper when filling haynets or mucking out and grooming; it will prevent hay, straw or hairs from sticking all over you.

126

Wear a stout pair of rubber gloves over woollen ones to keep your hands warm and dry when working around the yard in the winter.

127

If you tend to drop your whip when jumping or hacking, put an elastic band around the handle and slip your middle finger through it.

128

Glucose or alcohol is a good pick-me-up for tired or sick horses, or to encourage poor doers.

Both are readily absorbed into the blood-stream and are also palatable to most animals. A double handful of glucose may be mixed into a mash or a feed, or a bottle of stout added instead.

Stout is a good pick-me-up for tired or convalescing horses.

129

Feeding garlic can help to control a horse's worm burden, although it should not be used as a replacement for normal regular worming treatments. It can also help improve skin complaints and asthmatic conditions. Two or three cloves can be crushed and added to the feeds.

Garlic can be beneficial to horses.

130

Wet, muddy legs should never be washed with warm water, as this encourages the pores of the skin to open and allows dirt and bacteria to enter, starting up infections such as mud fever. Instead, use stable bandages over handfuls of clean dry straw. Leave them in place until the mud is dry and can be brushed off. If it is necessary to wash the legs, use cold water, and towel them dry afterwards.

131

To keep plaits looking neat and tidy at shows and to stop any wispy hairs from appearing during the day, try using setting lotion instead of water on the mane before putting each plait in.

132

Adding a double handful of glucose to feed or water will help a horse in shock to recover more quickly. It also helps to reduce the chances of 'breaking out' in a sweat again after a competition or hunting.

133

Snow can pack up into the hooves, forming little stilts, which may not only make your horse lose his footing but which can also strain the tendons. To help repel snow and ice spray a non-toxic silicon-based grease on to the soles of the hooves. Never smear used engine oil or grease on them, however, as this could cause damage.

134

Rather than buy best-quality vegetables to feed your horse, ask your local greengrocer if he has any which are past their best. He may sell them to you cheaply or even give them to you for nothing if they are not good enough for human consumption. So long as they are not actually mouldy, this is a cheap way of providing succulents for your horse, particularly in winter when grass is scarce. He will welcome the addition to his diet.

135

Mutton fat, saved from the Sunday lunch or begged from the butcher, is good for brittle feet. Simply rub it into them.

136

Wire fencing and electric fencing can be made more easily visible by tying strips of coloured plastic to it at regular intervals.

137

When putting hay out in the field, place the heaps in a sheltered spot so that they are not blown about. Better still, make a hay rack, which will stop the hay from being trodden into the ground and wasted.

138

A little talcum powder sprinkled into close-fitting long leather or rubber boots will make them easier to put on and take off.

139

A horse who tends to eat his straw bedding can be discouraged if you thoroughly mix in the old bedding with the fresh. A weak solution of Jeyes fluid sprinkled over the bed after it has been put down will also act as a deterrent.

140

Always use a quick-release knot when tying up a horse; in an emergency the free end can be pulled and will release instantly.

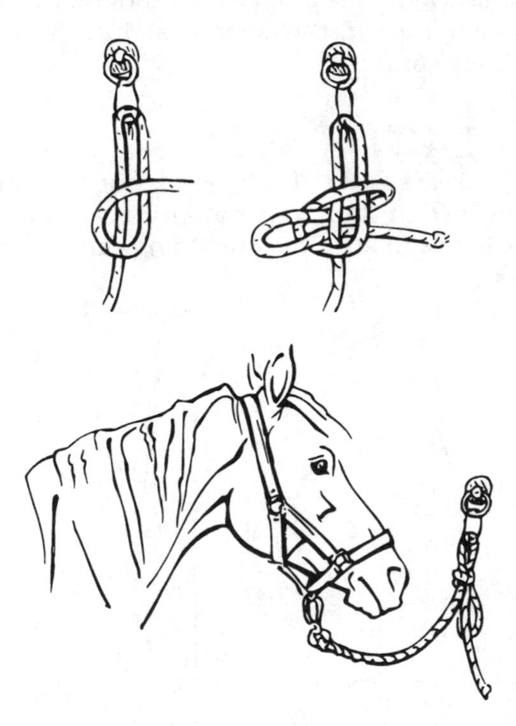

A quick-release knot.

141

When feeding hay in the field, always put out one pile of hay more than the number of horses – this will keep squabbling to a minimum.

142

Have two metal discs (of the dog tag variety) engraved with your name, address and telephone number, and attach one to your bridle and one to your saddle. Should you have an accident, causing your horse to run off, the person catching him will know who to contact.

143

Always make the effort to muck out and put up the bed in your trailer after use, so that it has a chance to air. Soiled bedding can rot the floorboards.

Muck out your trailer after use: wet bedding can rot the floorboards.

144

Weigh feeds rather than feeding by volume. Short feeds can be deceptive and may vary in weight from one sack to another. Similarly, sections of hay can vary tremendously in weight, and it is easy when feeding by eye either to underfeed or to waste money by overfeeding.

145

Freezemarking has proved to be an effective deterrent against thieves, and it is well worth protecting your horse in this way. Get together with a few friends, as there is normally a reduction in price for several horses.

146

Slice carrots lengthways before feeding, to reduce the risk of choking.

147

Build the bed right up to the doorway of the stable to help keep out draughts in winter. If the door does not meet the ground, attach a thick strip of rubber to the bottom of it to help keep draughts to a minimum.

148

When moving straw bedding to your horse's stable, put it into a sack – it will save you from dropping bits all over the yard and having to sweep them up later.

149

Keep a piece of baler-twine handy in your pocket when you go hacking; if your bridle breaks while you are out on a ride you can carry out emergency repairs.

150

If you do not have any boot trees, you can help leather boots to retain their shape by stuffing the legs with rolled up magazines. Alternatively, fill two old plastic lemonade bottles with water and place one in each boot, upside down. Either method will help the boots to retain their shape.

Old lemonade bottles make good boot trees.

151

If you are likely to be rushed for time during the week, fill up several haynets beforehand ready for use.

152

When grooming or clipping around the hindquarters of a ticklish horse, hold the tail firmly in your spare hand. If you hold it to one side he will be less inclined to kick out. If he does intend to kick, you will get an early warning by feeling his tail twitching when he becomes cross.

153

Use a summer sheet beneath stable rugs; they are far easier than rugs to wash and dry when they are dirty.

154

Exercising can be a problem during the winter when there is snow and ice. Construct a schooling ring of soiled straw or shavings from your horse's stable in a spare corner of the field. A 3 in to 4 in deep layer will provide you with a ridable track on which to walk, slow trot or lunge when the weather is really bad.

155

Avoid deep litter beds if your horse is prone to respiratory problems, as they can make them worse.

156

Wrap some insulating tape or Elastoplast around the handle of your whip with your name written on it in case you mislay it or someone else picks it up by mistake.

157

Quartermarks can be kept in place by using a quick squirt of hair spray – but be careful that the noise does not frighten your horse.

158

When taking photos of your horse, ask someone to squeeze a squeaky toy to attract his attention and encourage him to prick his ears.

159

Some horses who are bad to clip are upset more by the noise of the machine than anything else. If you think this is the case with your horse, stuff some cotton-wool into his ears.

160

If your horse's mane tends to lie on the wrong side, both sides, or, worst still, to stick up, pulling it from the underside will help. Then dampen it with water and plait it loosely, securing the ends of each plait with elastic bands. These stable plaits can be left in for a day or so, providing the horse does not rub his mane.

161

Stuff old plastic sacks with loose straw and use them as fillers beneath poles when jumping; they will make the fence more imposing, and mean that you need fewer poles in order to construct a course. Weight them down with stones to prevent the wind blowing them away, and staple the ends so that the stuffing does not come out.

Plastic bags stuffed with straw can be used as cheap, light jump fillers.

162

When turning a horse out in the field, leave the gate slightly ajar so that you can make a quick exit if necessary. Turn his head towards you before letting him go. If he kicks out or bucks on gaining his freedom, you are less likely to be on the receiving end.

163

Insulating tape or Elastoplast wrapped around the area of your lungeing whip where the thong joins the stock will protect it against wear and tear.

164

If your horse suffers from brittle feet, keep him on a peat bed in the stable. This will improve the condition.

165

Keep a record of your horse's pulse, temperature and respiration, so that you know what is normal for him and can check if you suspect that he is a bit off colour. Take a note of them twice a year, once during the summer, and again during the winter, since they will vary according to the season as well as the horse's age.

166

When walking a show jumping course in a competition, walk the jump-off course at the same time, in case you get through the first round. You will not have the opportunity later on.

167

Use only a discreet rosebud or similar for a buttonhole when entering showing classes. Pick it the night before, and keep it fresh and prevent the bud from opening by placing it overnight in the refrigerator.

Buttonholes (discreet ones only, please!) may be kept fresh overnight in the fridge.

168

If you have to bandage an injured leg, which needs support, or to hold a dressing in place, always bandage the opposite leg as well. It will probably be subject to more strain than usual, as the horse takes the weight off his injured limb.

169

If your horse has been off work for some time and you are just starting to get him fit again, harden up the skin in the saddle and girth areas with salty water or surgical spirit to reduce the chances of sores or galling.

170

If you discover that the drawstring of your haynet tends to merge in with the rest of it, plait up a new one from baler-twine of a contrasting colour.

171

When riding in cross-country competitions, hunting or show jumping, plait the bridle headpiece into the top plait, securing it well with thread; if you fall off, you will be less likely accidentally to remove the bridle.

172

Deep-litter beds should be completely cleared out every six months; the box should be disinfected, and wooden kicking boards creosoted to prevent their being rotted by the bedding.

173

If the keepers on your horse's bridle have stretched and always slide down, hold them in position by putting an elastic band on to the cheekpieces just beneath them. When the headpiece straps are placed through the keepers the elastic bands will hardly be noticeable and the bridle will look much tidier.

174

Space plaits evenly by cutting a piece of plastic comb to approximately 1 to $1^1/_2$ in in width. Use it to gauge the width of each plait accurately as well as to make straight divisions in the hairs between them.

175

Fussy feeders can be tempted by adding molasses or black treacle to the feeds, or sprinkling it over the hay from a watering can. Dilute it first in warm water in the proportions of one part treacle or molasses to five parts water. Because of their strong smell and taste, they can also be used to disguise the presence of worm powders or medicines.

176

Rubber bit guards can be fitted to bits easily by threading two pieces of twine through the hole in the centre. Tie each piece with a knot so that two loops are formed; hook one over a tack cleaning hook or door handle, and put your foot on to the other and push downwards with it. The rubber will stretch, the round hole becoming an elongated slot through which the bit ring can be passed. The twine can then be cut and removed.

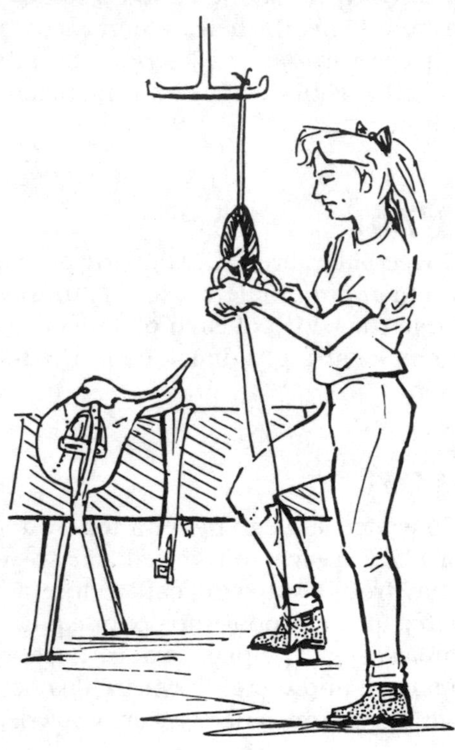

Use a piece of baler-twine to stretch rubber bit guards sufficiently for the bit rings to be pushed through.

177

If you use studs, keep the holes in the shoes clean and grit free when they are not in use by plugging them with a piece of lightly-oiled cotton wool twisted tightly in.

178

Rubber riding boots which have split around the ankles can be cut down to make useful short waterproof boots for use around the stable-yard.

Old rubber riding boots which have started to split around the ankles can be cut down to make short waterproof boots for use around the yard.

179

Pull the mane after exercise, when your horse is still warm, as the hairs will come out more easily when the pores are open.

180

If your horse or pony kicks at the stable door, nail a piece of old carpet, or a sack stuffed with straw, on to the inside. He will not injure himself, and the noise will be muffled.

181

If rugs start to rub along the shoulders or over the withers, check first of all that the rug fits as well as possible. Then pad these areas with real or synthetic sheepskin, and stitch in a foam rubber pad on either side of the withers. This will lift the rug a little, and also keep it from slipping to one side, which can cause rubbing.

Real or synthetic fleece stitched around the shoulders of rugs helps to prevent rubbing.

182

Rubber overreach boots can be made easier to pull on either by placing them in hot water for a few minutes so that they become more stretchy or by smearing a little Vaseline around the tops so that they slide on with less of a struggle.

183

Help to prevent thrush by using a limesock after picking out the feet. A limesock is made by putting a large handful of garden lime into the toe of an old sock and tying a knot in the ankle. Thump the sole of the horse's foot with the lime-filled end of the sock until it is coated with a thin layer of powder.

184

Paint your initials on your grooming kit so that it does not become mixed up with everyone else's. Use either gloss paint, nail varnish, or even the type of enamel used for painting plastic model kits.

185

Clean your hat ready for shows by brushing it gently with a soft brush to remove dust. Then steam it for a few minutes over the spout of a boiling kettle, and brush gently again to raise the nap and remove any marks.

186

When opening bales of hay or straw, turn the bales so that the knots of the twine are on top. Cut at the knots, take hold of them and pull. The twine will slide out easily, leaving you with lengths of twine which may come in handy in the future.

187

Metal dustbins make useful rat- and mouse-proof containers for feed, and will also help to protect it from damp.

188

A cheap poll guard for travelling can be made by cutting a slit at either end of a rectangular piece of foam rubber, and then slipping the headpiece of the headcollar through them.

A home-made poll guard.

189

If your horse is clipped at home, instal a power point on a central overhead beam in the stable. There will then be far less danger of his standing on the flex of mains-operated clippers.

190

Add a blue bag to the rinsing water when bathing grey or parti-coloured horses, to give a really blue-white finish.

191

If your horse feels the cold during the winter but has to live out, stitch an extra blanket into the lining of his New Zealand rug. It can easily be unpicked and removed again when the weather becomes warmer.

192

When going to shows, cut up a pair of tights, or use a stocking, to place over the tail. Bandage over both it and the dock to hold it in position. This will keep the tail clean during the journey.

193

Massage the back briskly after exercise to restore circulation completely – especially after long rides.

194

When measuring up your horse for a new rug, take the measurement from the centre of the chest (where the rug would fasten) along his side to where the rug would finish at the back. Keep him looking straight ahead, or the measurement will be incorrect.

195

Have winter rugs repaired in plenty of time before the cold, wet weather sets in. If you appear at the saddler's at the last moment, you will probably find yourself at the end of a long queue. Make sure they are clean too. If your saddler does agree to repair dirty rugs, he will probably charge extra for cleaning, since mud and grit can damage his machine.

196

If you are travelling your horse to a show in a trailer rather than hacking, keep his bridle clean by placing it inside an old pillow-case. Your own show clothes can be kept inside a dry-cleaning bag until you are ready to change into them. If you are hacking, wear track suit bottoms over your jodhpurs to keep them clean until you are due to go into the arena.

197

If you can afford it, buy a spare New Zealand rug; if one gets damaged or becomes soaked, you can then use the other.

198

When your horse is being clipped, put the bed up against the stable walls, leaving just a little down in the centre of the floor to give his hooves some grip. Take this scattering of bedding out afterwards and throw it away. Any hair or drops of clipper oil which become mixed in it could lead to digestive problems if he nibbles at the bedding later.

199

If you clip your own horse, use a piece of chalk to mark out the lines of your clip before you start, so that you do not end up with them lopsided.

To prevent mistakes, mark out clipping lines with chalk before you start.

200

Keep a string bin or bag near to your supply of hay or straw, and put all the pieces of baler-twine in it when you remove them from bales. If left lying on the ground they can easily become mixed up with loose hay or straw and end up in the bedding or in a haynet. If eaten they can lead to choking. By placing them safely out of the way, you will always know where to lay your hands on some twine when you need it.

201

A good way of stopping a horse from chewing the top of the stable door, and which also helps to discourage crib biters, is to cut a slit down the length of a piece of tough plastic drainpipe. It should then be slotted on to the top of the door and secured by a nail at each end. The curved, smooth surface will prevent the horse from being able to gain any purchase on it with its teeth.

Index